Table of Contents

Chapter One	- 1 -
Introduction	- 1 -
Chapter Two	- 4 -
History of Agile Project Management	- 4 -
A Detailed History of Agile:	- 5 -
Benefits and Rules of Agile Project Management	- 9 -
Chapter Three	- 18 -
Traditional To Agile Project Management	- 18 -
Traditional Project Management Model	- 22 -
Agile Project Management Model	- 25 -
Difference between ancient and Agile Project Management	- 27 -
Chapter Four	- 30 -
Why is agile project management necessary?	- 30 -
The Multiple and Varied Uses of Agile Project Management	- 30 -
Benefits of Agile Methodology	- 37 -
Key Aspects of Traditional Project Management Approaches and Agile Development	- 38 -
Chapter Five	- 41 -
Basic quality control	- 41 -
Agile Project Quality	- 44 -
Agile Quality Assurance and Control	- 45 -
Agile Quality Improvement	- 50 -
Chapter six	- 51 -
Tools or Methodology in Agile Project Management	- 51 -
Criteria For Choosing Agile Tools	- 51 -

Tools of the trade .... - 52 -
Source control tools .... - 53 -
Continuous integration tools .... - 54 -
Team management tools .... - 55 -
Chapter Seven .... - 94 -
Problem Solving Techniques .... - 94 -
Problem-Solving Strategies .... - 99 -
How to Solve Problems .... - 99 -
Agile Team Motivation .... - 102 -
Agile Failure Modes .... - 103 -
Chapter Eight .... - 105 -
Conclusion .... - 105 -

Agile Project Management:

An Easy Step by Step Handbook to Learn Agile Project Management and Make Innovative Products

By:
Alex Moore

© **Copyright 2019 - All rights reserved.**

The content contained within this book may not be reproduced, duplicated or transmitted without direct written permission from the author or the publisher.

Under no circumstances will any blame or legal responsibility be held against the publisher, or author, for any damages, reparation, or monetary loss due to the information contained within this book. Either directly or indirectly.

Legal Notice:

This book is copyright protected. This book is only for personal use. You cannot amend, distribute, sell, use, quote or paraphrase any part, or the content within this book, without the consent of the author or publisher.

Disclaimer Notice:

Please note the information contained within this document is for educational and entertainment purposes only. All effort has been executed to present accurate, up to date, and reliable, complete information. No warranties of any kind are declared or implied. Readers acknowledge that the author is not engaging in the rendering of legal, financial, medical or professional advice.

By reading this document, the reader agrees that under no circumstances is the author responsible for any losses, direct or indirect, which are incurred as a result of the use of information contained within this document, including, but not limited to, — errors, omissions, or inaccuracies.

Chapter One

Introduction

Project management is the application of information, skills, tools and techniques to project activities to satisfy the project necessities. in spite of industry, project management has been established to be an important part of a company's potency and its ultimate success. In fact, the organizations using tried project management practices waste twenty-eight less cash and implement projects that are 2.5 times a lot more flourishing. Project management professionals conclude that the definition of a successful project is one that is not solely completed on time and within budget, however one that conjointly delivers expected advantages.

For many years, there has been a traditional method of project management based on the idea that the customer can know and define his or her requirements fully up front. While this is occasionally the case, very often customers either don't know precisely what they want or, more often, "will know it when they see it." Agile project management is a repetitive and progressive approach to delivering necessities throughout the project life cycle. At the core, agile projects ought to exhibit central values and behaviours of trust, flexibility, authorization and collaboration. Agile project management focuses on continuous improvement, scope flexibility, team input, and delivering essential quality merchandise. Agile project management focuses on continuous improvement, scope flexibility, team input, and delivering essential quality merchandise. Agile project management approaches embody scrum as a framework, extreme programming for building in quality upfront, and lean thinking to eliminate waste.

Why does one want agile in project management? Agile could be a philosophy that concentrates on authorized individuals and their interactions and early and constant delivery valuable into an enterprise. Agile has enduring attractiveness and 'proved' itself in software system development. However, though the arguments are compelling, proof that it's additionally beneficial than other approaches remain for the most part anecdotal.

Firstly, a short summary of what agile project management is and the way it differs from a lot of ancient project management approaches.

There are many methodologies which will be accustomed manage an agile project; two of the simplest famed being scrum and Lean. an agile project's shaping characteristic is that it produces and delivers work in short bursts (or sprints) of anything up to a few weeks. These are repeated to refine the operating deliverable till it meets the client's needs.

Where traditional project management can establish an in depth set up and detailed needs at the beginning then decide to follow the plan, agile starts work with a rough idea of what's needed and by delivering something in an exceedingly short amount of time, clarifies the wants as the project progresses. These frequent unvaried processes are a core characteristic of an agile project and, as a result of this way of operating, cooperative relationships are established between stakeholders and also the team members delivering the work. Scope must be variable where no detailed needs exist initially, however agile still has processes to make sure that, at every stage, the work to be done is outlined and in-line with client wants. The role of project manager tends to be quite totally different on agile projects (and is usually referred to as the scrum Master or Project Facilitator); it's the team member who deals with issues and handles interruptions to permit the other team members to focus on producing the work.

So agile projects need documentation, reviews and processes even as traditional projects do to satisfy needs, manage prices and schedules, deliver advantages and avoid scope creep; agile merely doesn't place as much stress on extremely detailed documentation and does not expect to totally perceive the wants before work can begin. Instead it emphasises the importance of delivering a working product as one thing tangible for the client that can then be refined till it fulfils the client's wants. The key measure of project progress is this series of operating deliverables. there's clearly a risk to starting work on a project before the extent of that job is absolutely known however this risk is quenched by the speedy delivery of a working product, albeit one that's unlikely to be excellent initially.

Chapter Two

History of Agile Project Management

As against the standard methodologies, agile approach has been introduced as an effort to form software engineering versatile and economical. With 94 of the organizations practicing agile in 2015, it's become a typical of project management. The history of agile are often traced back to 1957: at that point Bernie Dimsdale, John von Neumann, Herb Jacobs, and Gerald Weinberg were exploiting progressive development techniques (which are currently called Agile), building software system for IBM and Motorola. Although, not knowing a way to classify the approach they were active, they all realized clearly that it had been completely different from the Waterfall in many ways. However, the modern agile approach was formally introduced in 2001, once a gaggle of seventeen software development professionals met to debate different project management methodologies. Having a transparent vision of the versatile, light-weight and team-oriented software development approach, they mapped it call at the manifesto for Agile software system Development. aimed toward "uncovering better ways in which of developing software", the manifesto clearly specifies the fundamental principles of the new approach:

A Detailed History of Agile:

Here could be a look into how Agile emerged, how it acquired the label Agile, and where it went from there. It's necessary to take a glance at where Agile software system development came from to get an understanding of where things are at nowadays.

Before 2001: A lot of individuals peg the beginning of Agile software system development, and to some extent Agile generally, to a gathering that occurred in 2001 when the term Agile software system development was coined. However, folks started operating in an Agile fashion before that 2001 meeting. beginning in the nineties, there have been numerous practitioners, either folks operating within organizations developing software products or consultants serving to organizations build software who thought, "You know what? The method we've been building software package simply isn't working for us. We've need to come back up with one thing completely different." These software developers started mixing old and new concepts, and once they found a combination that worked, they created a technique for his or her team to assist them bear in mind the combination of ideas that worked in an exceedingly given scenario. These methodologies emphasized close collaboration between the development team and business stakeholders; frequent delivery of business worth, tight, self-organizing teams; and sensible ways to craft, confirm, and deliver code. The people that created those methodologies patterned that others is also fascinated by obtaining a number of identical advantages they were experiencing, in order that they created frameworks to unfold the ideas to alternative groups in other organizations and contexts. this is often where frameworks like scrum, Extreme Programming, Feature-Driven Development, and Dynamic Systems Development method, among others, began to appear. The unfold of the ideas at this point was very organic, and every one of those completely different approaches began to grow in a very grassroots manner. folks borrowed the initial frameworks and tweaked them with completely different practices so as to create them applicable for his or her own contexts.

2001: There wasn't the same approach of describing these alternative ways to develop software package till a bunch of 17 individuals thought, "We're all doing these completely different approaches to developing software. we tend to need to get along and see wherever there are commonalities in what we're brooding about." The result was a gathering at a ski resort in Snowbird, Utah in 2001. When they got along, they did some sport and conjointly discussed where their approaches to software development had commonalities and variations. There were loads of things that they didn't agree upon, however there have been a number of things that they were able to agree upon, which ended up turning into the manifesto for Agile software Development. the 2 main things the Agile manifesto did was to supply a collection important statements that form the base for Agile software development and to coin the term Agile software development itself. In the months later on, the authors enlarged on the ideas of the Agile manifesto with the 12 Principles Behind the Agile manifesto. Some of the authors, together with Martin Fowler, Dave Thomas, Jim Highsmith, and Bob Martin, wrote up their recollections of writing the Agile manifesto. 16 of the 17 authors met at Agile2011 and shared their recollections of the event and their views on the state of Agile up to that purpose.

Post 2001: After the authors returned from Snowbird, Ward Cunningham posted the Agile manifesto, and later the 12 Principles. Agile Alliance was formally formed in late 2001 as an area for people that are developing software and helping others develop software explore and share ideas and experiences. Teams and organizations began to adopt Agile, led primarily by folks doing the development work in the groups. Gradually, managers of these groups conjointly started introducing Agile approaches in their organizations. As Agile became additional wide well-known, an ecosystem shaped that included the people that were doing Agile software package development and also the folks and organizations who helped them through consulting, training, frameworks, and tools. As the scheme began to grow and Agile ideas began to unfold, some adopters lost sight of the values and principles espoused within the manifesto and corresponding principles. rather than following an "agile" mindset, they instead began insisting that certain practices be done specifically in an exceedingly certain approach.

Organizations that focus entirely on the practices and also the rituals expertise difficulties operating in an Agile fashion. Organizations that are serious regarding living up to the Agile values and principles tend to comprehend the advantages they wanted and notice that operating in an Agile fashion isn't any longer one thing that's new and completely different. Instead, it merely becomes the approach they approach work. Agile Alliance continues to minister resources to assist you adopt Agile practices and improve your ability to develop computer code with lightsomeness. The Agile Alliance web site provides access to those resources together with videos and displays from our conferences, expertise reports, an Agile wordbook, a directory of area people teams and several other resources.

Benefits and Rules of Agile Project Management

Agile may well be a project delivery 'placebo'; operating as a result of those involved need it to. Agile empowers people; builds responsibility, encourages diversity of ideas, permits the early release of advantages, and promotes continuous improvement. It permits choices to be tested and rejected early with feedback loops providing advantages that don't seem to be as evident in waterfall.

In addition, it helps deliver modification once requirements are unsure, helps build client and user engagement by focuses on what's most useful, modifications are progressive improvements which might facilitate support cultural change. Agile will facilitate with deciding as feedback loops help economize, re-invest and realise quick wins.

However, Agile focuses on little progressive changes and therefore the challenge is that the larger image will become lost and create uncertainty amongst stakeholders. Building agreement takes time and challenges several norms and expectations. Resource value is higher; co-locating groups or invest in infrastructure for them to work together remotely. The onus is appeared to shift from the authorised end-user to the empowered project team with a risk that benefits are lost as a result of the project team is focussed on the incorrect things.

A critical governance decision is to select the appropriate approach as part of the project strategy. Level of certainty versus time to market is the balance that needs to be considered when selecting suitable projects to go agile. Organisations have to be realistic: the objective is not agile but good delivery, and a measured assessment of the preferred approach is essential to achieve that goal. This is defined by the project type, its objectives and its environment.

Agile is not a panacea, many practices its principles without knowing. Projects delivering end-user benefits is an agile principle which should also exist using traditional methodologies. Collaborative working will always: improve benefits; speed up delivery, improve quality, satisfy stakeholders and realise efficiencies.

There are 12 basic principles to successfully following an Agile project management development approach. At a ski resort in Snowbird, Utah, 17 software developers reflected on what defined the core principles of agile development methods. Their goal was to uncover better ways of delivering software and to help others do the same.

During that meeting, the Agile Manifesto was born. Comprised of 12 fundamentals, along with four core values, it provides the foundation of agile software development as we know it today.

- Attain customer satisfaction through continuous delivery of software
- Don't be afraid to make changes

- Deliver working software, with a preference to the shorter timescale
- Developers and management must work together
- Build projects around motivated individuals
- Face-to-face interactions are the most efficient & effective modes of communication
- Working software is the primary measure of progress
- Agile processes promote sustainable development
- Continuous attention to technical excellence and good design enhances agility
- Simplicity is essential
- The best architectures, requirements, and designs emerge from self-organizing teams
- Inspect & Adapt

Principle 1: Attain customer satisfaction through continuous delivery of software

Software is not built for the sake of building software. It's built to be put to use by an end user to better perform tasks that were previously out of reach, solve a problem, do their job better or more efficiently, etc. But often, the highest priority of software development is forgotten.

So, how can you better align with this principle?

Shorten the distance between requirements gathering and customer feedback by planning less change at a time. This gives you more opportunity to steer the software in a satisfactory direction for the customer.

Principle 2: Don't be afraid to make changes

You can implement changes now -- you don't need to wait for the next system to be built or a system redesign. Agile processes harness change for the customer's competitive advantage.

Shorten the distance between conceiving and implementing an important change. And even if it's late in the development process, don't be afraid to make a shift.

Principle 3: Deliver working software, with a preference to the shorter timescale

Previous development methods were front-loaded with tons of documentation under the guise of completing 100% of the requirements needed for a particular project. But towards the end of the project, the usual result was just that -- lots of documentation, but nothing to show for it.

Agile project management focuses on shortening the distance between planning and delivery. So, the agile methodology focusses more on creating software rather than just planning for it. This gives you the opportunity to improve the efficiency and effectiveness of the work.

Principle 4: Developers and management must work together

This one is crucial, especially because it doesn't come naturally to most people. Co-location between management and developers is usually the best way to handle this. You can also use communication tools for remote workers. It helps the two sides better understand each other and leads to more productive work.

Principle 5: Build projects around motivated individuals

There should be no micromanaging in agile project management. Teams should be self-directed and self-reliant. Make sure you have the proper team in place that you can trust to complete the project's objectives and provide the support and environment to get the job done.

Principle 6: Face-to-face interactions are the most efficient & effective modes of communication

Put simply, you want to shorten the time between a question and its answer. This is another reason why co-location or remote work during the same hours is key in agile project management. When teams work together under the same (virtual) roof, it's much easier to ask questions, make suggestions, and communicate.

Principle 7: Working software is the primary measure of progress

This is the primary metric an agile development team should be judged by: Is the software working correctly? Because if it's not, it doesn't matter how many words have been typed, bugs have been fixed, hours have been worked, etc. A good team needs to produce quality software -- all other measures are pretty much irrelevant if you can't get it working correctly.

Principle 8: Agile processes promote sustainable development

When working on the same project for a VERY long time, burnout can be a common problem among agile software development teams. To prevent this, work should be done in short productive bursts because excessive overtime cannot continue indefinitely without impacting the quality. Focus on choosing the right pace for the team members. Usually, the best pace is one that allows team members to leave the office tired yet satisfied.

Principle 9: Continuous attention to technical excellence and good design enhances agility

Developers shouldn't wait to clean up redundant or confusing code. Code should get better with each iteration. Along agile methodology, the software development team should use scrum tools and take time to review their solution. Doing this during the project saves you way more time than cleaning up code "later" -- which can also mean never.

Principle 10: Simplicity is essential

Keep things simple and minimize the time between comprehension and completion. Avoid doing things that don't matter -- such as the "busy work" that is so prevalent in corporate culture. Keep track of your team, count the hours worked in a fun way by using project management tools.

Principle 11: The best architectures, requirements, and designs emerge from self-organizing teams

A great agile management team takes its own direction. Members don't need to be told what needs to be done -- they attack problems, clear obstacles, and find solutions. It should be a red flag if the project manager has to micromanage.

Principle 12: Inspect & Adapt

This is a crucial principle in agile project management. At regular intervals, the team should reflect on how to become more effective, tune and adjust its behaviour accordingly. If there is a better way of moving a project forward, the team should implement adjustments.

Chapter Three

Traditional To Agile Project Management

In recent years, Agile methodology has become popular with several software package development groups as a result of the enhanced potency it brings regarding. quite a few corporations were unable to style and make desired products within optimum time and price thanks to their use of traditional project management methodology. By going agile they were able to fully remodel their processes and alter the means groups view project management.

Project Management: Project management is the discipline of initiating, planning, executing, controlling, and closing the work of a team to attain specific goals and meet specific success criteria. Regardless of the scope, any project ought to follow a sequence of actions to be controlled and managed. in keeping with the Project Management Institute, a typical project management method includes the subsequent phases:

1. Initiation;
2. Planning;
3. Execution;
4. Performance/Monitoring;
5. Project close.

Used as a roadmap to accomplish specific tasks, these phases define the project management lifecycle. Yet, this structure is just too general. A project typically incorporates a variety of internal stages within each phase. they can vary greatly relying on the scope of work, the team, the business and therefore the project itself.

In attempts to search out a universal approach to managing any project, humanity has developed a major number of Project management techniques. based on the above-described classic framework, traditional methodologies take a bit-by-bit approach to the project execution.

Thus, the project goes through the initiation, planning, execution, monitoring straight to its closure in consecutive stages. typically referred to as linear, this approach includes a number of internal phases that are successive and executed in a chronological order. Applied most typically to the development or manufacturing industry, where very little or no changes are needed at each stage, traditional project management has found its application within the software engineering still. referred to as a waterfall model, it's been a dominant software development methodology since the early 1970s.

Traditional (or waterfall) project management, majority of the times, follows a set sequence

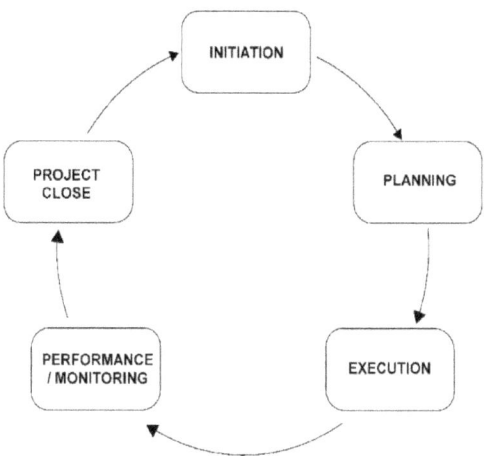

When one process is complete, only then can the next one begins. Typically, this method is more suited for projects that anticipate very few changes from the start to finish. Like manufacturing a billion-dollar Heli carrier. Here requirements are fixed, only cost and time vary. Since there are minimal changes, chances of the budget or time estimation going haywire are comparatively less.

However, not every project can be planned in the same manner. For software development projects, frequent iterations are required, a flexibility which Agile methodology provides. Instead of planning the entire project beforehand, teams focus on quicker iterations and increase efficiency. Although most of the steps involved remain same, it is not necessary that they are carried out in a sequential manner. These steps are broken down into smaller segments known as sprints.

Both Traditional and Agile Project Management have their merits and demerits. It depends on the nature of the product and the circumstances around it which determine the type of method to be used. It is wise to first understand the difference between the two before you consider adopting either one.

Traditional Project Management Model

Traditional project management is a universal practice which includes a set of developed techniques used for planning, estimating, and controlling activities. The aim of those techniques is to reach the desired result on time, within budget, and in accordance with specifications. Traditional project management is mainly used on projects where activities are completed in a sequence and there are rarely any changes.

Traditional Project Management or waterfall model includes a robust stress on planning and specifications development: it's considered to take up to 40oth of the project time and budget. Another fundamental principle of this approach is a strict order of the project phases.

Waterfall could be a linear, consecutive design approach where progress flows downwards in one direction—like a waterfall. Originating in the manufacturing and construction industries, its lack of flexibility in design changes within the earlier stages of the development process is because of it turning into exuberantly costlier due to its structured physical environments.

The methodology was first introduced in a piece of writing written in 1970 by Winston W. Royce (although the term 'Waterfall' wasn't used), and emphasizes that you're solely able to move onto the next part of development once the current phase has been completed. The phases are followed in the following order:

1. Specification requirements

2. Design
3. Implementation
4. Verification
5. Maintenance

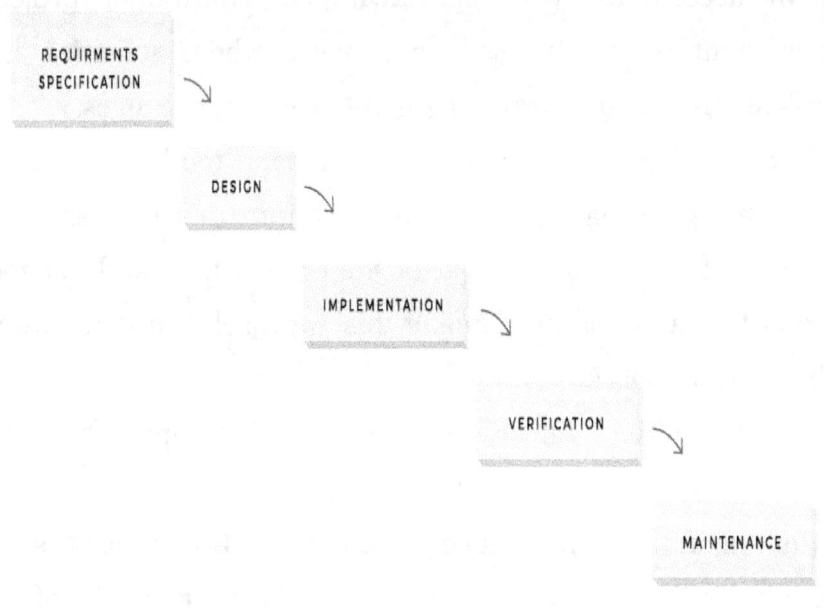

Waterfall may be a project management methodology that stresses the importance of documentation. the thought is that if a employee was to depart during the development process, their replacement will begin wherever they left off by familiarising themselves with the data provided on the documents.

Pre-Agile saw the waterfall methodology being employed for software package development, however there have been several problems because of its non-adaptive design constraints, the dearth of client feedback obtainable throughout the development process, and a delayed testing period. Best fitted to Larger projects that need maintaining rigorous stages and deadlines, or projects that have been done numerous times over where possibilities of surprises throughout the development method are comparatively low.

A new project stage doesn't begin until the previous one is finished. the method works well for clearly defined projects with one deliverable and fixed deadline. waterfall approach needs thorough planning, extensive project documentation and a tight management over the development process.

In theory, this could result in on-time, on-budget delivery, low project risks, and sure final results. However, once applied to the particular software engineering process, waterfall methodology tends to be slow, pricey and inflexible because of the various restrictions. In several cases, its inability to regulate the product to the evolving market necessities usually ends up in a large waste of resources and eventual project failure.

Agile Project Management Model

As opposed to the traditional methodologies, agile approach has been introduced as a trial to make software engineering flexible and efficient. With 94 of the organizations practicing agile in 2015, it's become a typical of project management.

The history of agile can be traced back to 1957: at that time Bernie Dimsdale, John von Neumann, Herb Jacobs, and Gerald Weinberg were using incremental development techniques (which are currently referred to as Agile), building software for IBM and Motorola. Although, not knowing a way to classify the approach they were practicing, all of them realized clearly that it absolutely was totally different from the waterfall in many ways.

However, the contemporary agile approach was formally introduced in 2001, once a bunch of 17 software development professionals met to debate various project management methodologies. Having a transparent vision of the flexible, light-weight and team-oriented software development approach, they mapped it out in the manifesto for Agile software Development. aimed toward "uncovering higher ways of developing software", the manifesto clearly specifies the basic principles of the new approach:

Through this work we have come to value: individuals and interactions over processes and tools operating software over comprehensive documentation client collaboration over contract negotiation Responding to change over following an idea.

Complemented with the Twelve Principles of Agile software, the philosophy has return to be a universal and economical new way to manage projects

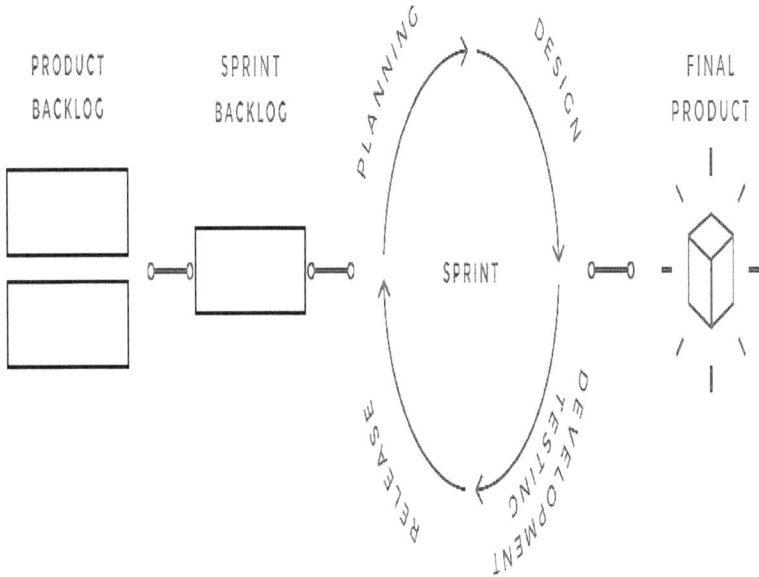

Agile methodologies take a repetitive approach to software development. in contrast to an easy linear waterfall model, agile projects include variety of smaller cycles - sprints. every one of them could be a project in miniature: it's a backlog and consists of design, implementation, testing and deployment stages among the pre-defined scope of work.

Difference between ancient and Agile Project Management

- **Flexibility:**

Traditional project management provides very little to no scope for creating changes to the product. It's a rigid process that solely follows a top-down approach. Once the plan is finalised, managers communicate it to their groups and check that that everybody sticks to that within the absolute best manner. there's plenty of resistance to any modification that's projected because it will disrupt the project schedule.

Agile methodology is a lot of adaptable and offers plenty of flexibility in terms of creating changes to the product. It permits team members to experiment and find out some of the simplest alternatives. they're free to communicate any plan they believe can facilitate to enhance the product additional. Being a feature-based approach, Agile focuses a lot of on obtaining the correct product than follow rigid structures.

- **possession and Transparency:**

In traditional management, possession belongs to the Project Manager. it's the manager's responsibility to arrange and document the whole journey of the product. except managers, solely customers are concerned within the planning stage however once implementation begins their involvement is zero. Since managers hold all the reins of the project, team members typically don't have a say in the results of their efforts or however the project is progressing.

Whereas in Agile methodology, the team members share possession of the project. everybody puts their heads along to come back up with a plan designed to complete the work within the calculable time and price. they're ready to read the progress of the product right from the beginning to its finish. Such transparency plays a very important role in maintaining a productive and extremely engaged work atmosphere.

- **problem solving:**

In case of surprising obstacles, people have to be compelled to escalate the difficulty to their managers. However, approaching your manager each single time isn't a possible possibility. It will cause undue delays and exceed the calculable time limit, except increasing the overall value additionally.

Agile groups have the authority to take choices on their own. they fight to internally solve all problems to avoid dalliance. Being closely concerned in the process, their data helps them to tackle most of the issues that hinder their progress. Unless there's a desire to require extreme choices, team members seldom need to escalate trivial matters to their manager.

- **Checkpoints and monitoring progress:**

Traditional method advocates heavy planning at the analysis and design stage of the project. Their focus is a lot of on streamlining the processes than on the product itself. Once the process is finalised, it's expected that the team can follow it step by step with negligible guidance. Progress is decided when the project is completed. There are not any frequent check-ins unless the manager receives any escalations.

Since there are shorter and faster iterations, Agile methodology encourages team members to have checkpoints at regular intervals. it's straightforward to work out the progress additionally as helps people maintain accountability in their work. In scrum – one in every of the foremost well-liked agile methodologies, groups hold daily stand-ups to catch up on what was the work done yesterday, agenda for the day and if there are any obstacles.

In the dynamic atmosphere that we have a tendency to currently live in, there are only a few things where modification doesn't occur. In such times, sticking to Agile methodology can serve companies far better.

Chapter Four

Why is agile project management necessary?

Agile project management has its disadvantages such as less easy identification of project risks and poor management of resources, and many project teams don't understand how to use agile project management effectively. However, with the fast pace of business change in the 21st century many projects need to be sure they will deliver something that meets client needs at the end of the project and not expend wasted effort refining requirements that will be out of date by the time the end-product is delivered.

Even in business environments that do not change rapidly it can be difficult to fully articulate requirements without seeing a tangible product first so there is still the risk of delivering something that doesn't quite meet the client's needs. That is why agile is becoming increasingly necessary for many different types of projects.

The Multiple and Varied Uses of Agile Project Management

It's no secret that the Agile method is quick turning into the preferred way to manage projects. within the year ahead, Agile is predicted to become even additionally widely used. Organizations, groups and even project management software are more and more responding to a requirement for more accommodative and evolutionary processes. And for good reason. in an exceedingly fast-changing business world that must answer speedy market and technology shifts, Agile delivers. Agile is ideal for any project that needs a series of versions or iterations that require to be reviewed and improved on until the final product is prepared for prime time. for instance, rather than waiting six months for a deliverable that's either blemished or now not meeting the current necessities, Agile permits you to manufacture a primary draft within as very little as two weeks (or less) for immediate feedback—and from here you'll improve upon every version till it's complete.

There are plenty of additional advantages to exploitation Agile. Here are thirteen reasons why groups like yours are exploitation this versatile project management method more and more:

• Agile is evolutionary, giving groups a chance to be told with every new iteration or draft.

• Agile lets groups deliver a prototype and improve upon it with each cycle.

• groups will manage shifting priorities a lot more effectively.

• This quick and versatile method will increase productivity.

• Agile supports regular and collaborative troubleshooting.

• The inherent collaborative nature of Agile improves project visibility.

- Agile helps groups and people effectively prioritize work and options.
- groups will anticipate incoming project changes.
- groups will build quick-course corrections based on stakeholder feedback.
- groups can prototype a solution or process for the consecutive version of the project.
- Stakeholders and clients will give feedback because the project evolves—without holding the project up (because the feedback is a component of the process).
- groups get speedy feedback from every version or iteration.
- Empowers project groups to work creatively and effectively.

Agile methodologies address completely customer's wants. throughout the total cycle, user involvement is inspired, providing visibility & transparency, showing the particular progress of projects. As mentioned earlier, Agile method is all about unvarying planning, making it very simple to adapt once some necessities change (if you're employed within the software development trade, i'm positive you recognize what proportion they'll change!). the actual fact that there's continuous planning and feedback through the process means we tend to begin delivering business price from the start of the project. Again, the thought is to deliver business price early within the process, making it easier to lower risks related to development. Let's go through the main advantages of agile project management, point by point.

1. High product quality

In Agile development, testing is integrated during the cycle, which suggests that there are regular checkups to ascertain that the product is functioning during the development. this allows the product owner to create changes if required and therefore the team is aware if there are any problems.

- defining and elaborating requirements just in time so the information of the product features is as relevant as doable.
- Incorporating continuous integration and daily testing into the development process, permitting the development team to deal with problems whereas they're still recent.
- Taking advantage of machine-driven testing tools.
- Conducting sprint retrospectives, permitting the scrum team to incessantly improve processes and work.
- finishing work using the definition of done: developed, tested, integrated, and documented.
- software is developed in progressive, speedy cycles. This ends up in tiny progressive releases with every release building on previous functionality. every release is totally tested to make sure software quality is maintained.

2. Higher client satisfaction

The product owner is often involved, the progress of development has high visibility and flexibility to change is extremely vital. this means engagement and client satisfaction.

- Demonstrating working functionalities to customers in each sprint review.
- Delivering products to market faster and additional typically with each release. The clients get early access to the product during the life cycle.
- Keeping customers involved and engaged throughout projects.

3. increased project control

- Sprint conferences.
- Transparency.
- Jira usage (visibility of each step of the project for each parties).

4. Reduced risks

- Agile methodologies just about eliminate the probabilities of absolute project failure.
- always having a working product, beginning with the very initial sprint, in order that no agile project fails fully.
- Developing in sprints, making certain a brief time between initial project investment and either failing quick or knowing that a product or an approach can work.
- Generating revenue early with self-funding projects, permitting organisations to pay for a project with very little up-front expense.

• Agile provides freedom once new changes have to be compelled to be enforced. they'll be enforced at little or no price due to the frequency of latest increments that are made.

• Adaptation to the client's desires and preferences through the development process. Agile commonly uses user stories with business-focused acceptance criteria to outline product features. By focusing features on the wants of real customers, every feature incrementally delivers worth, not just an IT component. This conjointly provides the chance to beta test software once every iteration, gaining valuable feedback early within the project and providing the power to create changes as needed.

5. Faster ROI

The fact that agile development is unvarying means the options are delivered incrementally, thus advantages are completed early whereas the product is in development process.

• Development starts early.

• A functional 'ready to market' product after few iterations.

• initial Mover Advantage.

• Long delivery cycles are usually a haul for businesses, particularly those in fast-moving markets.

• Agile suggests that quick product releases and skill to measure customer reaction and alter accordingly, keeping you ahead of the competition.

• focusing on Business worth. By permitting the client to work out the priority of features, the team understands what's most significant to the client's business, and may deliver options in the most beneficial order.

Benefits of Agile Methodology

With Agile project management, primary constraints, like time and price, are often unceasingly evaluated. speedy feedback, continuous adaptation and Q&A best practices are designed into teams' schedules, which ensures quality output and a streamlined process. the subsequent represents a number of the other advantages of Agile project management:

- increased focus on the particular desires of clients
- Reduced waste through minimizing resources
- increased flexibility enabling groups to simply adapt to change
- better management of projects
- quicker project turnaround times
- quicker detection of product problems or defects
- increased frequency of collaboration and feedback
- Improved development process
- increased success as a result of efforts are more focused
- speedy deployment of solutions

-

Key Aspects of Traditional Project Management Approaches and Agile Development

Traditional Project Management Tools	Agile Project Management Tools
Rigid team members must adapt	Flexible. Can adapt to team members
Abstract	Visual and intuitive
Works best with large teams	Works best with small and medium teams
Reporting is an element of major importance	Constant feedback is important
Works best with teams that share a work space	Can adapt to remote teams
Interest in final result	Interest in each sprint release
Well established quaterly or annual meetings	Daily meetings
Schedule a series of event	Schedule release and product versioning
Works with critical path calculations	Works with burn down charts
Formal management structure	Informal management structure
Dependencies between tasks can be dealt with on the go	Dependencies between tasks must be dealt with as soon as

	possible as they directly affect product releases
Quality Within	
Traditional Quality Management	**Agile Development**
Focused on delivering project objectives	Focused on constant improvement of delivered products
Adapts to established requirements	Adapts to customer changing demands
Comprehensive documentations is mandatory	Working software is more important than documentation
Progress is monitored through reports and periodical meetings	Progress is monitored through daily meetings and results
Defines working products criteria	Works with user stories
Reactive response to change	Proactive response to change
Sustainable development	Sustainable development
Complex solutions	Simplicity is essential
Considers that best decisions are made by professionals	Allows teams to self-organize and gives freedom to team members in choosing architecture, requirements and design

| Gives power to team leader | Gives power to team members |

Chapter Five

Basic quality control

The history of the word quality encapsulates plenty of definitions and aspects as well as case studies. As of lately, quality in all its forms has become a major factor taken into account by all companies and small businesses as well. In this sense, and in order to help companies integrate quality into their everyday business lives, standards, procedures, tools and techniques come to their aid to ensure a standardization and ease of implementation. An interesting aspect emerges however from observing the correlations between the implementation and maintenance of quality throughout a product's lifecycle and the project management process of that product. Another important aspect to factor in is the way that the agile approach and quality merge together to ensure quality products and/or services are delivered.

Quality Management, in a project context, is concerned with having the right processes to ensure both quality product and a quality project. This article describes Traditional Quality Management, Agile versus Traditional Quality Management, Agile Product Quality, Agile Project Quality, Agile Product Quality, Agile Quality Assurance and Control, and Agile Quality Improvement. Quality management is a method for ensuring that all the activities necessary to design, develop and implement a product or service are effective and efficient with respect to the system and its performance. Quality management can be considered to have three main components: quality control, quality assurance and quality improvement. Quality management is focused not only on product quality, but also the means to achieve it.

There are a lot of terms which include the word quality so here is the basic definition of related quality items

Quality policy: An organization's general statement of its beliefs about quality, how quality will come about and its expected result.

Quality management: The application of a quality management system in managing a process to achieve maximum customer satisfaction at the lowest overall cost to the organization while continuing to improve the process.

Quality management system: A formalized system that documents the structure, responsibilities and procedures required to achieve effective quality management.

Quality assurance: All the planned and systematic activities implemented within the quality system that can be demonstrated to provide confidence that a product or service will fulfil requirements for quality.

Quality control: The operational techniques and activities used to fulfil requirements for quality.

Quality plan: A document or set of documents that describe the standards, quality practices, resources and processes pertinent to a specific product, service or project.

Quality audit: A systematic, independent examination and review to determine whether quality activities and related results comply with plans and whether these plans are implemented effectively and are suitable to achieve the objectives.

Project Quality: Notice that Quality Management is concerned with both Product Quality and "the means to achieve it" which Project Perfect: Project Quality Planning calls Project Quality. Project quality is the "things like applying proper project management practices to cost, time, resources, communication etc. It covers managing changes within the project". Interestingly this is pretty close to what is covered by Agile Project Management.

Quality Assurance: Crosby made the analogy that quality assurance is like a person possessing a driver's license. Possessing the driver's licenses provides some confidence that the person is a safe driver

The emphasis of traditional quality assurance is producing a quality plan. A good quality plan, like a driver's license, offers confidence that quality will result. Project Perfect: Project Quality Planning suggests that in a project context the plan should answer the following questions:

1. What needs to go through a quality check?
2. What is the most appropriate way to check the quality?
3. When should it be carried out?
4. Who should be involved?
5. What "Quality Materials" should be used?

Other key aspects of quality assurance is producing the quality materials themselves. These include standards, guidelines, checklists, templates, procedures, process, user guides, example documents, and the methodology.

Quality Control: Continuing Crosby's analogy, if quality assurance is like a person possessing a driver's license, then quality control is actually checking that that person is a safe driver. In a project context quality control is about implementing the quality plan, i.e. doing the quality checks described in the plan. Normally just doing the check isn't sufficient and there needs to be proof the quality check took place. In other words the results need to be documented somehow.

Agile Project Quality

Agile is very much concerned about product quality in the sense of "Fitness for use" rather than "conformance to requirements". The first of the Principles Behind the Agile Manifesto is: Our highest priority is to satisfy the customer through early and continuous delivery of valuable software.

It manages to include both satisfied customers and a valuable product, both very quality oriented aspirations. I believe the other element of this principle, continuous delivery, is the key feature of Agile and this principle is very clearly tying continuous delivery to customer satisfaction. Project quality includes "things like applying proper project management practices to cost, time, resources, communication etc. It covers managing changes within the project"

Agile sometimes suffers because people confuse a poor implementation of the method with a limitation of the method itself. A poor implementation will leave out certain key Agile Project Management practices and quality will suffer.

Agile Quality Assurance and Control

Quality Assurance is planning activities to demonstrate quality and Quality Control is implementing those plans. To assure quality a traditional project management project manager would, if they were being thorough, produce a quality plan for the project. Agile project managers don't do this because the Agile process itself provides the quality assurance and control. Agile builds quality in to the product through a combination of practices from Agile Project Monitoring and Control and Agile Project Execution.

Product Owner in the Team: The team is trying to build software that meet's the Product Owner's intent, rather than what they wrote down, so in the Agile world the Product Owner becomes part of the team and guides development. Sometimes the Product Owner cannot devote 100% of their time to the project, but this is a risk that traditional projects also face.

Releasable Software Every Timebox: Each Timebox is meant to result in releasable code. Although meeting the Product Owner's expectations is a priority it is not the only criteria. Releasable software:

1.   Meets the Product Owner's expectations given features asked for and the Timeboxes completed so far

2. Meets agreed coding standards

3. Has to best design for the currently implemented features (via refactoring)

4. Is easily maintainable (via refactoring)

5. Has been tested to the satisfaction of the team and relevant stakeholders

Unscheduled Product Reviews: Because the Product Owner is part of team they get opportunities to informally review the product. Despite the fact these reviews are informal, teams are encouraged to document the results. Assign a scribe, take notes, type them up, and email them out. I always send the email to the Product Owner and Copy To the Technical Lead and other members of the team working in that area. This is part of Agile Change Management.

Scheduled Product Reviews: The product is formally reviewed in the Timebox Review Meeting at the end of each Timebox. You can schedule more formal reviews during the Timebox, and for longer Timeboxes (say 3 or 4 weeks) this is a good idea. DSDM mandates the results of these reviews are documented. I think this is a good idea but I'm not too tied to review documents. It is recommended to document the reviews in the same as was described in the Unscheduled Product Reviews, i.e. an email to the relevant stakeholders.

Frequent Status Meetings: The team monitors project progress in the Daily Team Meeting. It is also an opportunity to review whether the team is following the agree approach.

Automated Unit Tests: Automated unit tests express the internal behaviours of the software. The total suite of automated unit tests become regression tests and are used to verify that the internals of the software continue to function as designed after subsequent changes. Agile teams aim for 100% pass rate for automated unit tests. More mature teams write the tests before they write the code in Unit Test Driven Development.

Acceptance Tests: In DSDM there are documented quality criteria for all work products, but looser forms of Agile restrict this to the User Stories. User Stories are the high level description of the external behaviours and business rules of your software. Each User Story has at least one acceptance test. The acceptance tests elaborate the brief description provided by the User Story. They define the scope of the story and clarify the Product Owner's intent with concrete examples. This clarifies the Product Owner's intent, points the team in the right direction, and confirm when the intent has been met.

Where possible Acceptance Tests should be automated. As with the automated unit tests, the suite of automated acceptance tests become regression tests, validating that the customer's intent continues to be met by the software after each change to the code. You won't automate all Acceptance Tests – it won't be possible to automate some and others won't be cost-effective to automate. But if you want the test repeated as part of a Regression Test then it is better to automate. If you can write a test so that a person can repeat the steps consistently then you can probably write a automated test and let the computer repeat the steps even more consistently.

More mature teams write the tests before they write the code in Acceptance Test Driven Development.

Test Driven Development: In Test Driven Development the tests become the specification. Because the tests are automated there is no ambiguity, the software either passes the test or it fails.

Regression Testing: A regression test is the repeat of an earlier test. Usually that means Unit Tests and Acceptance tests. Regression tests ensure that changes to the software have not broken good code. My experience is that if regression tests are manual they don't happen. It is with regression testing that the real value of test automation is shown.

Exploratory Testing: Exploratory Testing uses un-scripted tests to quickly identifying new types of problems with the software. Show stoppers (such as system crashes or unhandled errors) are usually fixed immediately. Less serious problems might be deferred. This testing might also reveal new User Stories to be scheduled into later Timeboxes.

Specialist Testing: Extra testing activities like performance testing are scheduled in the same way as User Stories.

Code Review: The Two Pairs of Eyes approach provides a peer review of code to check it follows agreed coding standards, conforms to design guidelines and is easily understood by developers other than the author. This can be either through pair programming or more traditional code reviews/walkthroughs.

Code Metrics: Although not mandated by any of the Agile approaches, some Agile teams, like some traditional teams, collect metrics about the quality of the code. Examples are code-coverage (amount of code covered by unit tests), conformance to maintainability design principles (e.g. Lack of Cohesion of Services, Normalised Distance from the Main Sequence), and language-specific metrics.

Continuous Integration: Continuous Integration is about maintaining quality all the time, throughout the project. It involves automatically integrating and running a regression test every time somebody does a check in. This is likely to happen several times a day. Running an automated regression test frequently means defects are highlighted soon after they are introduced (i.e. when the build goes Red, i.e. fails). The team's top priority is to get the build Green again.

Informative Workspaces: The Project's Informative Workspace is the primarily place to show data on Project status. Typically it has Burn Down Charts, the Timebox Plan, perhaps the current build status, and anything else that might be a particular concern at the time (e.g. quality metrics such as test coverage). Essentially it is a way to monitor product and project quality on a daily basis.

Scheduled Project Reviews: The Retrospective part of the Timebox Review Meeting is a chance to review the project as whole. Problems are addressed as User Stories in subsequent Timeboxes.

Agile Quality Improvement

One of the Principles Behind the Agile Manifesto is: At regular intervals, the team reflects on how to become more effective, then tunes and adjusts its behavior accordingly.

Retrospectives are the mechanism most Agile teams use for reflection. During a Retrospective the team looks at how well the Timebox went and what they can do different. The high priority changes become User Stories to go into the Release Plan for implementation. Often this is the process to implement new Agile processes.

Chapter six

Tools or Methodology in Agile Project Management

Every agile company has agile project management tools to follow the methodology in a more efficient and strict manner. If you're still doubting which one to choose or if you would like to know which are the leading ones on the market this article is definitely the one for you.

Criteria For Choosing Agile Tools

The best agile tools supply the following most important elements for agile project management. Look at elements outside of their feature set, such as their user interface, their usability (how easy is it to learn?). Also evaluate how much value the tool offers for the price—how its price stacks up against other tools with similar features and functionality.

In terms of features, look for the following in evaluating the best agile tools in this review:

Task management – Kanban or Scrum boards with projects, task lists and everything else that goes with it – from files and discussions to time records and expenses.

Team collaboration – Communicate updates with local and distributed teams, and share task lists, feedback, and assignments

Agile metrics, reporting & analytics – Time tracking and projection, easy-to-understand progress reports for stakeholders, quality assurance, and progress with tools to identify and remedy project obstacles, evaluate performance, and appraise financials

And finally, check for integrations. Ensure that the tool plays well with the right tools. In the case of agile tools, which are often used for developing software, treat integrations with software development and issue management tools with higher priority. However, keep in mind teams in non-development environments won't need this type of integration and would benefit more from integrations with other work apps like Slack, Google Apps, Adobe, etc.

Tools of the trade

The key to success in agile development is to enable flexibility while maintaining organization. The best way to do this is to deploy a set of good tools that help track the project and organize the team's progress. They don't impose strict schedules and roles, but merely make it easier for the developers to self-manage and converge on their goals.

There are dozens of software products designed to help managers set priorities and developers write code that addresses them. Some of these tools are designed to track different forms of development, including projects that are more centrally managed, but they are flexible enough to be used for agile development. Others are built specifically to fit the agile model and nurture as much programmer freedom as possible.

The tools support the project by helping the team identify the requirements and split them into a number of smaller tasks. Then it tracks the programmers as they work collaboratively on the parts. The process is often split up into short cycles that gradually converge on the final result. The cycles alternate between planning sessions and code sprints. Keeping the cycle short and including plenty of developer feedback in the planning lets the team adjust and focus.

A common feature of all these agile tools is a graphic dashboard that reports how the team is progressing and meeting the goals. Some of the more sophisticated tools are integrated with code repositories and continuous integration tools that automatically graph how the new code is evolving. Is the latest code passing tests? Are more features coming online? These questions are all answered on a dashboard that everyone can see. When the team can follow each other's progress visually, they're better able to stay on track.

Another important part of this process is communication. Good agile tools organize the discussion and planning. The developers can focus on each of the features, tasks, or bugs in separate threads. Splitting the discussions up helps the project move forward at the right rate for each section.

Here are the top tools that are forming the foundation teams rely upon to ship code on time or even ahead of schedule.

Source control tools

Git, like some of the other tools here, wasn't built just for agile teams but is still essential. It offers much of the flexibility that teams need to move ahead. The lack of one dominant central repository makes it simpler for different developers to follow different paths and then merge their code later. Git is widely supported, and many teams now use its hosting services to keep their code organized. Many of the other tools in this list take their cues from Git and use the updates to the repository to track and test progress. Other top source control tools include Mercurial, Subversion, and CVS.

Continuous integration tools

Just like Git, continuous integration tools aren't explicitly designed to support agile development, but it would be hard to imagine running a large agile team without their help. The tools automatically add a layer of processing when code is committed, helping to ensure that the team is working smoothly together. The tools have hundreds of plugins for tasks such as creating documentation or compiling statistics. Their most important job is running unit tests that ensure the software is performing correctly after all the new code is added to the stack. Many of the tools in this list also use the results from post-commit testing to determine how quickly the code is meeting goals.

There are a number of good continuous integration tools that play well with agile management systems. Some of the best known tools include Hudson, Jenkins, Travis CI, Strider, and Integrity.

Team management tools

Agile Manager

HP's Agile Manager is built to organize and guide teams from the beginning as they plan and deploy working code through the agile model. At the early stages of the cycle during the release plan, the managers gather the user stories and decide how the teams will attack them. These set the stage for the sprints and deployment.

During each code sprint, the scrum masters and developers record their progress on the user stories and issues. All the progress (or failures) from the build and the unit tests are plotted in charts on a dashboard so the entire team can watch how they're converging on the release.

The tool gathers information directly from major tools such as Jenkins, Git, Bamboo, and Eclipse. To complete the cycle, Agile Manager will push stories and tasks directly to these tools so developers can keep track directly from their favorite IDE.

Active Collab

From juggling tasks to tracking time and generating bills, Active Collab is organized to help software shops deliver code and account for their time. The heart of the system is a list of tasks that can be assigned and tracked from conception to completion. A system-wide calendar helps the team understand and follow everyone's roles. The system checks the amount of time devoted to all the tasks so the team can determine how accurate their estimates are.

The system also supports a collaborative writing tool so everyone can work together on documentation, an essential operation that sets the stage for more agile collaboration later.

The tool can be hosted locally or used through a cloud service.

JIRA Agile

The JIRA Agile tool adds a layer for agile project management that interacts with the other major tools from Atlassian. The team creates a list of project tasks with a tool called Confluence and then tracks them on an interactive Kanban board that developers can update as they work. The Kanban boards become the center of everyone's focus in planning how to attack the code. JIRA is a tool developed for bug and issue tracking and project management to software and mobile development processes. The JIRA dashboard has many useful functions & features which are able to handle different issues easily. Some of its key features and issues are: issue types, workflows, screens, fields and issue attributes. Some of these features you won't find elsewhere. The dashboard on JIRA can be customized to match your business processes.

The Agile tool is well-integrated with other Atlassian tools. The dashboard updates the moment code is committed to Stash or Bitbucket, Atlassian's Git hosting products. Bamboo (see number three above) offers continuous integration that builds and tests the code before reporting the relative success or failure back to the main JIRA page. Discussions take place through HipChat, which indexes the discussions to the tasks.

Agile Bench

The Agile Bench tool is a hosted platform that emphasizes tracking the work assigned to each individual. The release schedule begins as a backlog of user stories and other enhancements. As they're assigned, the team must gauge both the business impact and the cost of development by assigning an estimate of the complexity of each task in points. The dashboard tracks both of these values so that members can tell who is overloaded and which tasks are the most important. Agile aims to deliver valuable software through close collaboration between all members of the team and stakeholders, welcoming changes to requirements and frequent software releases. This approach means working software can be used much sooner, which reduced confusion around what was actually being delivered. It also emphasises the use of motivated and talented teams of individuals who focus on technical excellence and good design to enhance their agility

The tool is well-integrated with standard Git hosting sites like GitHub or Bitbucket (see number five above), allowing it to make committed code with tasks. If your project needs more, there's also an open API that can integrate the project information with any other system. Popular Alternatives to Agile Bench: Slack, SimplifyEm Property Management, Basecamp, Asana, ClickUp, WorkZone Project Management, Hubstaff, Smartsheet, EclipsePPM, monday.co.

Pivotal Tracker

Pivotal Tracker is a project-planning tool for software development teams. It will help to visualize your projects in the form of stories or virtual cards, break down projects into manageable chunks, have conversations with clients about deliverables and scope. Tracker can divide stories into future iterations, learning from a team natural pace of work. It can accurately predict the estimations and project's completion. Tracker has a transparent team view of priorities to help each member's objectives. Pivotal Tracker encourages a practical agile software development process.

Pivotal Tracker is a straightforward project-planning tool that helps software development teams form realistic expectations about when work might be completed based on the team's ongoing performance. Tracker visualizes your projects in the form of stories (virtual cards) moving through your workflow, encouraging you to break down projects into manageable chunks and have important conversations about deliverables and scope. As your team estimates and prioritizes those stories, Tracker divides them into future iterations, learning from your team's natural pace of work to accurately predict when you will complete future work. Tracker's transparent team view of priorities means that everyone knows what needs to be done, what is being done, and when it will be completed. Tracker's agile philosophy not only helps your team keep pace and plan work, but adjust and change course when the unexpected happens, so your team can deliver earlier and more consistently.

Pivotal Tracker is just one of a constellation of tools from Pivotal Labs created to support agile development. The core of the project is a page that lists the tasks that are often expressed as stories.

Team members can rank the complexity with points, and the tool will track how many tasks are being finished each day. The constellation includes Whiteboard for team-wide discussions, Project Monitor for displaying the status of the build, and Sprout, a configuration tool.

Telerik TeamPulse

Teampulse is an on-premise agile project management software that aims to improve software development processes. In Agile project management, Teampulse enables users to manage requiremnts and bugs. It also hel[s in planning the release of products and in tracking work progress while keeping the project team in cosntant communication and collaboration. It does so through a set of intuitive features that enhances efficiency.

Telerik is known for its numerous frameworks for creating apps for the mobile marketplace. They've bundled much of that experience from creating their own code into TeamPulse, a tool they use to track projects. The main screen displays a page full of tasks that need to be completed and follows the team as it progresses. The menus offer configuration options and a wide variety of reports showing how the project is evolving toward completion. It also works with Telerik's other tools for building and testing code. Telerik allows you to improve our process from requiremnts gathering through planning, managing and monitoring, resulting in smoother deliveries.

VersionOne

VersionOne is a formidable Agile management solution that is both comprehensive and versatile and developed for teams and projects of various scope and size. It is a compact platform that delivers outstanding performance in terms of managing and tracking multiple teams, tasks, and projects. Simple to use and highly scalable, VersionOne continues to excel in the Agile market, highly used and recommended by businesses of any shape or size.

VersionOne is structured to support other agile software methodologies, including Kanban, Hybrid, Scrum, Lean, SAF, and XP enabling companies to scale agile quicker, simpler, and smarter. Right from planning your portfolio and program level initiatives to tracking and offering value to your customers, the software empowers delivery and reduces time to market. Having an end-to-end visibility of your project's progress and performance equips you with comprehensive insights needed to make data-driven decisions for new plans, changes, and potential issues.

When a large enterprise embraces agile development, they need a tool that's customized to juggle multiple teams working on multiple initiatives because eventually they'll need to work together. VersionOne is designed to organize all the groups involved in development across an enterprise by providing a stable communication platform where everyone can plan the initiatives and create persistent documentation.

The tool embraces Kanban boards for following ideas and stories through the process until they're turned into working code. The system tracks all sprints and organizes the retrospective analysis so the team can start the cycle again.

Additionally, the openAgile API makes it possible to integrate Version One with other packages.

Planbox

Planbox is the pioneering provider of cloud-based AI-Powered Agile Innovation Management solutions – from creative ideas to winning projects. Our mission is to help organizations thrive by transforming the culture of agile work, continuous innovation, and creativity across the entire organization. Our family of products includes Collaborative Innovation Management, Team Decision Making, and Work Management applications. Planbox is designed to provide agile innovation tools for everyone, built for companies and teams of all sizes. Planbox is the comprehensive innovation solution trusted by some of the world's most recognized brands, including Blue Cross, Cargill, Caterpillar, Dow Chemical, Exxon Mobil, Honeywell, John Deere, Novartis, Ontario Power Generation, Sun Life Financial, Whirlpool and Verizon, with millions of internal and external users.

Planbox offers four levels of organizational power to keep multiple teams working together toward a common goal. At the top are initiatives, which are the biggest and broadest abstraction. They contain projects, which are built on items that, in turn, are filled with tasks. As the team finishes the tasks, Planbox tracks the progress on all these levels and produces reports for all stakeholders. One clever feature lets you loop in customers so they can voice their opinion before the code is set in stone. The time tracking feature lets everyone compare the time they spend on an item with the estimate of how long it was thought to take. The tool integrates with Github (see number six above) for code storage, Zendesk for tracking customer satisfaction, UserVoice for bug tracking, and many more.

LeanKit

LeanKit is a cloud-based visual management tool based on a Kanban-style platform that caters to businesses of all sizes across various industry verticals and helps them to implement lean principles, practices and work methodologies across business functions. LeanKit allows organizations to connect project boards at the team and project level and provides users with project visibility. Users can assess project status and manage project dependencies. It also helps users to visualize workflow process and receive real-time updates about project activities and the status of different tasks.

LeanKit also features a reporting and analytics module with metrics such as flow, quality, throughput and lead time. Users can generate custom reports that help them to spot trends and make business decisions. The solution supports integration with various third-party systems such as JIRA, Pivotal Tracker, Salesforce, Zendesk and more.

LeanKit is a visual project delivery tool that enables teams of all types and across all levels of the organization to apply Lean management principles to their work. LeanKit aims to imitate the conference room whiteboards where most projects begin. It lets all team members post virtual notes or cards that represent all the tasks, user stories, or bugs that must be addressed. As the team finishes them, the board updates faster than any whiteboard. The software also allows multiple teams to work together in separate spaces while still coordinating their interactions.

Axosoft

Another widely used Agile Project Management software solution that can be used for bug tracking is Axosoft. Mainly, it is used by software and application developers that are keen of Scrum framework. It has a rich set of tools that every developer needs to ensure that they create and deliver fully functional, bug-free software on schedule. Axosoft's project tool tracks the project in three different ways. The Release Planner offers a tabular view of the different tasks, bugs, and user stories. Developers drag and drop the different entries to assign them and mark them as finished. The burndown charts show graphically how quickly the team is converging on its goal. The projected ship date is displayed prominently to keep everyone on track. The planning is also done Kanban-style using the card view, where each card represents one task.

With Axosoft, developers can create viable plans for development, plot the steps of the process, collaborate effectively and seamlessly, identify issues and resolve them on time prior to delivery. Everything is centralized, ensuring transparency and that everyone is on the same page; supporting the feedback and dialogues option with a customer. Planning with Axosoft should be more effective as the software platform allows to gather all the details and specific information to create the right product backlog. This definitely makes the planning process easy, from creating the steps, scheduling the release, managing the versions and sprints all the way to completion. With the Daily Scrum Mode, project managers, Scrum Masters and other members of the team can see who is assigned to what task and how is one progressing. One useful feature is the customer portal that makes it possible for customers to weigh in on the development process by requesting features, giving feedback on designs, or testing new code.

Agilean

Agilean is an AI and NLP-based SaaS Enterprise workflow automation and management solution that caters specifically to small and medium IT companies. Agilean helps set and automate your Kanban processes under a couple of minutes from the ground up from a selection of over fifty inbuilt templates. Agilean is fully customizable and straightforward for ease of use. Agilean is designed to streamline workflows but also enhance existing assignments based on the specifications of the organization or client. It accelerates and improves user capabilities for project planning, execution, monitoring, control and continuous learning for a variety of software and other industry vertical projects.

Agilean Features include Single dashboard for all projects, visual and clear overview of tasks, drag and drop tasks between columns easily, limit work in progress to be more efficient, horizontal Swimlanes, import Boards, tasks, subtasks, attachments, and comments, automated actions, Gantt Charts, backlog management, create, edit, and move Cards, file attachments, work item breakdown, multi-team work distribution, custom fields, lead and cycle time, work distribution, assigned user, cumulative flow and burn-up, burndown, advanced reporting, work-in-process (WIP) limits, WIP violations and override, map the process, identify impediments, perform the average risk of the project, share response plan and strategy, monitor, control, and close the impediments, schedule meetings and set agendas and simplify meeting minutes, automated follow up on all action points, and real-time collaboration.

With Agilean, the workflow can be constantly improved by removing bottlenecks and time-wasting processes. The dashboard allows a clear view that enables efficiency and a simple way to keep clients happy by responding quickly to their needs and concerns.

It is a SaaS enterprise workflow automation and project management software solution that is basically created to be used by small-medium IT enterprises. The main features of Agilean include project planning, execution, monitor, impediments and response plan, stand up meeting automation, release management, retrospective analysis, and visualized reports.

Wrike

Another great one from the list of the best agile project management tools is Wrike that is one of the best in terms of integrating email with project management, having main features inside. It is built to scale and drive results by giving you the flexibility you might need to manage multiple projects and teams at one place. Along the Agile process, you going to get the accurate, up-to-date information and you can insert & add any important information inside. Planning should be easier, there will be always accurate information and real-time reports and analytics that going to save your time and help in analysing the situation. This is for sure one of the resources that your team might do like to use daily: customization supportive & collaboration tools and many other things that will keep your team focused. The flexibility provided by Wrike enables multifunctional groups to collaborate and get things done effectively from a single location. The service allows you to schedule, prioritize, discuss, and keep track of both work and progress in real time — all with just a few clicks of the mouse.

Wrike has been the project management software of choice for many Fortune 500 companies, such as, Google, Stanford University, Adobe, HTC, and EA Sports to mention a few.

Trello

I guess many of you already know about Trello, one of the most used and well-known project management applications. It has both free and premium accounts that give you a great chance to use most of the common functions. The structure of Trello is based on the Kanban methodology. All the projects are represented by boards, that contain lists. Every list has progressive cards that you make as drag-and-drop. Users that are related to the board can be assigned to said cards. Also, it has many nice, small but not less useful features I would like to indicate: writing comments, inserting attachments, notes, due dates, checklists, coloured labels, integration with other apps, etc. Additionally, Trello is supported by all mobile platforms. What I also like about Trello is that this tool can be used both for work, like we do it in Apiumhub, and personal processes. Trello has a variety of work and personal uses including real estate management, software project management, school bulletin boards, lesson planning, accounting, web design, gaming and law office case management. A rich API as well as email-in capability enables integration with enterprise systems, or with cloud-based integration services like IFTTT and Zapier.

Kanbanize

Kanbanize is a Kanban software for Agile project management that brings full transparency within both individual team workflows as well as across the entire organization. The tool is successfully adopted by a number of industries including Product Development, IT Operations, Marketing and Advertising, Legal and Financial services, etc. Kanbanize is the go-to solution for teams and companies looking to better organize their work, manage multiple projects, track progress and make their work processes more efficient. The software supports highly customizable Kanban boards that allow you to adapt to frequently changing requirements. There, consumers can use timelines to plan their initiatives with greater agility, break them down into manageable tasks, create multiple workflows for cross-functional teams and track overall aging work in progress. The system also supports a powerful analytics module that allows project managers to measure different types of metrics such as lead time, cycle time, team throughput, etc., so they can continuously improve their work processes and make them more predictable.

Backlog

Backlog is an all-in-one online project management tool for task management, version control, and bug tracking. With features like sub tasking, custom issue fields, and Gantt charts, it's easy for teams to define, organize, and track their work. Burndown charts, Git & SVN repositories, and Wikis help developers review, track, release, and document their code. And targeted notifications keep everyone in the loop along the way. Bringing together the organizational benefits of project management with the power and convenience of code management, Backlog enhances team collaboration across organizations large and small.

Assembla

Assembla is a set of tools and services developed to speed up software development and provide support for distributed agile teams. There are 2 platforms offered by Assembla to ensure that teams have the tools and the capabilities to manage, deliver, and maintain not just apps and Agile projects, but websites too. These 2 Assembla products are Assembla Workplaces and Assembla Portfolios. Assembla Workplaces combines various tools and build them around a team list or social activity stream. These include code repositories, management, ticketing and issue management, and collaboration. Assembla Portfolio gives users total control over multiple projects and Team Workspaces. The product comes with a centralized user management feature and reporting plus a branded portal.

Assembla is a combination of cloud-based tasks and code management tools for software developers. The aim of Assembla to move development teams from the typical Scrum agile toward something that is more continuous, distributed, and scalable. Assembla is a provider of Apache Subversion hosting along Git, P4, Dropbox integration, agile task management, team collaboration and project management. With this tool, you can cover all aspects of a project, from ideation to production, as well as to upload large media files, manage code reviews, document your work. Also, if you have any apps you would like to integrate you will easily do it, for example, with Github or Slack.

Asana

Asana is one of the most popular project management software currently available on the market. The robust work management platform serves your teams so they can stay focused on the goals, projects, and daily tasks as you grow your business. To get your works organized, Asana enables you to plan and structure work in a way that's best for you. It handily lets you set priorities and deadlines, share details and assign tasks—all in one place. To stay on track, it allows you to follow projects and tasks through every stage. You know where work stands and can keep everyone aligned on goals.

To help you meet deadlines, the platform lets you create visual project plans to see how every step maps out over time. With it it's much easier to pinpoint risks and eliminate roadblocks. Even when plans change.

Asana is the ultimate task management tool. It allows teams to share, plan, organize, and track the progress of the tasks that each member is working on. It is simple, easy in usage and is free for up to 30 users in a team. As all the previous agile project management software platforms with the main objective in allowing us to manage projects and tasks. What is noticeable is that you don't need to have even an email to use Asana. Each team can create its workplace that will contain projects and project's tasks; each task can have notes, comments, attachments and tags.

This tool can be used as for the small processes and for the giant ones without any limits in the industries or departments.

Binfire

Binfire is an online project management and collaboration tool for decentralized and large teams operating in multiple locations. It helps virtual teams to plan, monitor, and coordinate several projects simultaneously, using a common workspace. Binfire offers all features needed by teams in a single location, so that all files and tasks related to the project can be accessed easily through this one application. Thus, Binfire creates a virtual office space that improves collaboration and communication in the team.

In addition, Binfire is a well-integrated and moderately priced platform, with enterprise pricing adjusted to the needs of different businesses and industries. A 30-days free trial is also available for interested companies to explore the features, and decide which plan works the best for them. Nevertheless, prospective users can count on the company's experienced team to guide them through the process and help them choose, as they can be contacted via phone, email, or 24/7 live support directly on their website. Binfire supports all major project management methodologies including Agile, Waterfall and Hybrid Project Management. It provides real-time collaboration with such features like an interactive whiteboard, message board, burndown charts, project folders, collaborative PDF mark-up, real-time notifications, status updates and much more. In the task management, you can find issue management, bug tracking and document collaboration sections.

Drag

Drag transforms your Gmail into organized Task Lists. It's a free Chrome extension that turns your inbox into a manageable workspace (just like Trello, but for Gmail). Now, many well-known companies like Uber, Airbnb, Netflix, Spotify and others use it to improve the efficiency of email management. On the 18th June they are launching Drag Team! Now, not only individuals can organize their inbox in Trello-style boards, but also teams can collaborate on emails, right from inside Gmail. This means that teams working together in a project or managing accounts such as sales or support don't need to manage multiple external apps to get things done.

Proggio

Proggio is a project management software that puts a premium on collaboration and teamwork. It is built to handle all sorts of projects, both long-term (highly integrated projects) and short-term (continuous stream of deliveries), simple and complicated. With Proggio, project teams have a clear view of their tasks, schedules, and priorities. Managers can easily plan their steps, monitor their progress, and effectively manage budget, resources, and manpower. The project dashboard can be customized to provide users with all the information they need at a single glance.

Users who are on the go can easily see what's going on, what needs to be done, and make critical decisions as Proggio lets them view their project status, tasks, and more from a mobile device. Proggio brings clarity and innovation to agile project management. Proggio is based on a holistic approach to project management, placing people in the center, not tasks. It drives success through creating a shared purpose, building momentum and staying focused. The application introduces several awesome features: amazing project plan visualization, team collaboration in one easy click, patented automatic analysis and process improvements.

nTask

nTask is a cloud-based task management solution that caters to small businesses and individuals. It provides users with tools that enable collaboration with team members, task management, meeting scheduling and more.

With nTask, users can assign tasks, generate progress reports, set recurring tasks, share files, attach files to tasks and generate checklists. Gantt Charts help users monitor project schedules. The solution also enables users to plan and monitor budgets for different projects, allot resources, define risks and issues and monitor team members' time spent on different tasks. From making checklists to managing projects, collaborating with project teams, scheduling meetings, sharing files and more, nTask lets you do everything using just one tool. Spend less time managing tasks and more time doing things with a simple and easy-to-use task board. Worth trying!

OneDesk

OneDesk is designed to improve team collaboration and encourage participation from third parties, and effortlessly connects developers and clients to a productive environment of cooperation for product development. Installed with a rich set of features, OneDesk can do a lot of things to help users come up with better and more customer-centred products and services, from gathering valuable feedbacks and innovative ideas, delivering top class customer support, to monitoring social media for mentions, customer trends, and current buzz, whether positive or negative

With OneDesk you can manage your projects, support your customers, provide services. This tool makes it easy to make successful projects and deliver them on time. It combines Agile and traditional project management: Gantt charts, scheduling, assignments, discussions, notifications on tasks & issues, time-tracking with timesheets & task timers, reporting, exporting, plan releases, roadmaps and many more!

VivifyScrum

VivifyScrum is an Agile project management tool which can be used both by small teams and large organizations. For collaboration, teams can choose a customizable Scrum board which enables working in Sprints, as well as Product Backlog management and other Scrum practices. Teams that prefer Kanban can choose to collaborate on a Kanban board. High-powered item cards allow for a clear description of tasks, sharing all pertinent files and establishing relations between different tasks. In addition to this, VivifyScrum provides features for managing multiple projects across the organization. Thanks to advanced team management functionalities, a company can ensure that the right people work on the right projects. The tool also has an inbuilt time tracker that automatically creates worklogs for team members. Companies can also invoice their clients straight out of VivifyScrum.

VivifyScrum is a cloud-based agile project management solution that features Scrum and Kanban collaboration boards, team management, invoicing, client management and time management. It is suitable for small agile teams and large organizations in multiple industries. Users can create their virtual organizations and add their team members. After creating projects, it is possible to assign various roles to team members and add their engagements on projects. Projects can be linked to the relevant collaboration boards, where team members can track their work. The Scrum board offers product backlog, multiple active sprints, sprint goal, burndown chart and stats per user.

StoriesOnBoard

Stories Onboard is a visual planning tool for Product Owners, agile teams, based on the "user story mapping" method. It's integrated to JIRA, Trello, GitHub, Azure DevOps, Pivotal Tracker etc. More than 1500 companies use this tool worldwide.

StoriesOnBoard is a tool where you can break down your ambitious goals into tangible pieces. Then you can create a roadmap for reaching your goals by identifying the tasks that move toward them the most. By creating a story map, you will be able to see the big picture any time, thus instead of losing in tiny details you can focus on your goals for reaching them in a timely fashion. You can share you story maps with your remote team members so you will be able to work with them online.

Nuvro

Nuvro is a full-featured project management platform that provides the necessary tools to small online businesses at an affordable price range. It empowers owners and managers with a bird's eye view of the projects and tasks their teams are handling with a smart and visual workload calendar. Necessary details such as deadlines and overdue tasks are also visible from this. As such, leaders can act on issues appropriately and with haste.

Moreover, Nuvro promotes smooth collaboration between teammates. It has a file sharing solution with no limits, a team inbox, and a notes module that allows sharing. Because of this, your staff can help each other do better at their assigned responsibilities as well as complete tasks and projects faster. Efficient project execution often relies on many tasks, subtasks and sometimes sub-subtasks. Nuvro was engineered to make it easy to quickly create and assign these tasks to suitable team members. In Nuvro everything's transparent and everyone's accountable for their share of every project. Spend less time trying to bend current software to meet your needs and more time accomplishing your business goals.

Orangescrum

Orangescrum is a task management, collaboration, and project management software combined, providing project managers and teams with an effective platform to help them perform their functions efficiently and improve productivity. The software helps in centralizing all your projects and tasks as well as managing your resources, from people, processes, and technology so that you are able to finish your project in time and without compromising its quality. Orangescrum is also equipped with time tracking capability to help monitor your team members and how much time they spent on their tasks. It also provides you with real-time analytics on all areas of your projects as well as your members, delivering you a clear picture of your projects and enabling you to identify areas that require your concern and attention.

Orange Scrum has a cloud, cloud self-hosted, and open source enterprise editions. Teams can work anywhere and anytime with its Android and iOS mobile apps.

Orangescrum is a simple and effective Project and Task Management software with Open source and Cloud editions. It allows for Agile Scrum Project Management with intuitive Scrum Board, Sprints, Epics, Story points and Velocity Charts. Shape and execute your projects with robust task management (task groups, sprints, tasks & Subtask & Kanban Boards). Interactive Gantt chart allows for real time task progress monitoring and dependency mapping. Orangescrum with its integrated time, resource & invoicing mgmt. prevents the use of multiple tools thereby helping your teams to stay productive and organized. Real time executive reports dashboard makes up for quick informed decisions by your executives. Thus, Orangescrum has it all to put you in control of your business and stay ahead of your competition. It is widely preferred by Small, Medium and Enterprise users. Orangescrum has been ranked among top Project Management Software list for the year 2018 by Accurate Reviews and awarded with Great User Experience and Rising Star for 2017 by Finance online and Compare camp among others.

Zoho Sprints

Zoho Sprints is an online agile project management solution designed to help agile teams plan their project, track their progress, and deliver the appropriate product on time. The simple and clutter-free tool takes care of keeping timesheets, monitoring the task statuses, preparing meetings, and overviewing the analytics. Given that agile teams function with a core value on responding to change and working with a sense of urgency, having a system that augments this type of operation is vital. Zoho Sprints is a dynamic software that is quick and easy to set up, so you can immediately invite your team members, assign them roles, build a backlog, and commence your sprint.

It is part of the revolutionary suite of software solutions by Zoho designed to help businesses in various aspects of their operations. Hence, you can easily integrate Zoho Sprints with Zoho's other applications in marketing, sales, accounting, customer support, and more using a single login and password.

Zoho Sprint is a clutter-free planning and tracking tool for agile teams. It allows you to create user stories, add estimation points, stay on track with personalized scrum boards, and schedule your review and retrospective meetings from one place. Aside from the "To do", "In progress" and "Done" columns, the Zoho Sprint's Scrum Board will allow you to add your own custom columns so they match your team's unique needs, and by using Timesheets you can give your clients a picture of the time involver or estimate your next sprint more properly. Zoho Sprint can be used as an app for both iOS and Android devices, and it allows you to sign up for free.

ProofHub

ProofHub is a cloud – hosted project management solution that helps you to stay on top of deliverables and deadlines. It has scalable features and pay terms that match the requirements of any business size from small start-ups to large enterprise. The Walnut, CA-based company has attracted big ticket clients, such as, TripAdvisor, Harvard University, and Wipro, since the software was launched in 2011. Likewise, ProofHub was among the Cloudswave Awards 2014 top ten project management software for excellent features and performance. Recently, the company introduced ProofHub Bolt apps for Android and iOS users, which made their project management service available even out of the office.

The software helps managers across the critical phases of a project: from planning to organizing, to managing and delivering outcomes on time. It's a central hub for teams, clients, and contractors to share notes, tasks, knowledge, and discussions for a more efficient collaboration and timely response.

A single platform brings together managers and decision makers to discuss plans, create notes and to-do lists, lay down Gantt charts, and calendar milestones and daily tasks. They can drill down the requirements and deadlines to their respective team members by sharing notes, files, schedules, and timesheets. A

Proofing tool also helps relevant parties to discuss, comment, and finalize a document or file in a single window. This tool does away with the cumbersome practice of shuffling email messages and attachments back and forth to finalize a file like logo designs, floor plans, or site map.

Meantime, the Gantt charts, milestone calendar, timesheet, and reports help managers and team leaders track the project and measure on-going performance. Any issues, obstacles, or bottlenecks are easily identified for quick resolutions. ProofHub also features advanced functions, including: Casper mode privacy control; custom domain; SSL encryption, IP restriction, and daily backup; custom roles; and advanced search.

The software integrates well with popular collaboration and productivity tools such as Google Drive and Dropbox and email servers to enhance the overall collaboration experience. Moreover, ProofHub's web browser platform is compatible with any devices and popular operating systems, including iOS, Android, and Windows.

With ProofHub's work management tool you also gain control with Gantt charts to create the project plans that'll ensure timely completion of the projects, assign Project manager to keep a bird's eye view on project progress, create Custom roles to define access levels for the teams and clients, and Project reports for insights on how the projects are progressing, and how teams and individuals are performing.

GanttPRO

GanttPRO is an online project management solution that utilizes a Gantt Chart approach to help users become more efficient and productive in managing their projects, from the conceptualization phase all the way to the realization and delivery. The software is built upon the combined years of experience and top-grade project management expertise that culminate into a powerful platform that integrates all essential project management tools as well as maximum ease-of-use.

GanttPRO is created with the purpose of simplifying workflows, improving collaboration and communication, and delivering projects within accurate estimates. It makes tracking the progress of your projects a breeze and enables you to share your Gantt charts easily with your colleagues and clients during your presentations, reports or business plans.

GanttPRO is one of the most affordable and robust online Gantt chart software in the market with a very intuitive interface. For task and project planning, it offers a visually appealing Gantt chart timeline where all the tasks, dates, and deadlines are clearly seen. This tool is great for team collaboration. You can describe your requirements, add links to useful resources, write comments, and attach files. If any change in a project occurs, GanttPRO notifies team members immediately with the help of push-notifications and emails. For an easy start, GanttPRO offers ready-made templates that will be useful for professionals from different spheres.

Teambook

Teambook is an efficient project management and collaboration tool especially targeted at small and mid-sized businesses, as well as freelancers and service consultants. Its easy-to-use planner with straight-out-of-the-box functions can at once turn your team leaders into effective project schedulers and managers. You can direct company resources for specific tasks using the simple planner. Likewise, you can see who's working on which activities in real-time and track performances and deliverables with the help of analytics. Using tags or filters to refine metrics, search the fitting resource for specific activities.

Teambook also helps you keep tab of the teams with individual resource homepage, email alerts, and iCal link that syncs with popular calendar apps. You client is also updated immediately with a dedicated link to your project milestones. Furthermore, Teambook integrates with popular tools, such as: Harvest, Zapier, Google Calendar, iCal, and Outlook. You can also use its API to integrate your own application

Many organizations don't need a complex and expensive all-in-one project management tool to manage their team. That's where Teambook comes up as a simple and affordable alternative. Teambook is a dedicated capacity planning and resource scheduling online tool that focuses on improving a team's productivity by taking better planning decisions. Its simple visual planner is designed to ensure a smooth planning experience. It gives a fast overview of your team's occupation, adding a booking (unique or recurring) is a matter of seconds and you can edit or move various bookings at once with the bulk editing mode. Plus, Teambook brings powerful filter and custom tag systems to make sure you can easily allocate the right, available resource to the right task. Also, you can identify productivity issues through the simple reporting feature that gives an overview of your resources' availability, productivity and utilization.

Chapter Seven

Problem Solving Techniques

Agile development focuses majorly on adding value to customers. Agile team work on iteration and at the end of every iteration, customer feedback is evaluated which paves the way for further improvement. All these working on iteration and regular interval changes might create host of problems in agile project management. To resolve these problems there are host of problem solving techniques that needs to be taken care of by the project manager and the agile team. Stopping a problem as soon as recovered is an important recipe of success for any project, be it agile or not. While looking for an agile problem, the agile team working on the project needs to look at different levels.

Process level: In this level, the agile team needs to recognize how they are doing with agile adoption.

Quality and performance level: Based on quality and performance level, the project manager needs to understand individual skills and should think of ideas to do better.

The team dynamics dimension: The agile team needs to understand the requirements to make the agile team work better towards achievement of the goal.

There are various Agile Project solving techniques that can be followed by agile team in order to resolve problems in agile and work effectively towards achievement of the agile project. Few of these agile project solving techniques are discussed below.

Establish yourself as a devil's advocate: Make sure that as a project manager or a leader, you will ask a lot of questions. The questions should be asked in order to help the entire team towards some progressive work along with opening up to alternate ideas.

Be kind, rewind: Ask a lot of questions to your agile team and push them towards solving the problems and at the same time motivate them to generate new ideas.

Don't let them waste more time on meta-problems: Relentlessly, focus on the big problem and not on the smaller or Meta problems. It is tempting to stay on Meta problem as they are smaller but not always helpful.

Ask probing questions: Scrutinize the team and understand what they are thinking by asking a lot of probing questions.

Use reflective listening: Reflective listening is a technique in which the listener repeats or paraphrases what he has just heard. The purpose of this technique is to confirm understanding of customer's feedback and let them know that they are heard.

Avoid injecting your own ideas: Even if you have a better idea, always let your team speak first and try to pick the best idea out of them. This will help your team to become more interactive and motivate them to put forwards their ideas.

Lead them to the answer: Always make a habit of allowing your agile team towards the answer. Even though you know the answer, avoid giving it to them directly. Rather ask questions and give hints and lead them towards the answer.

These are some of the best agile problem solving techniques followed by agile practitioners. To know more about agile problem solving techniques and agile project management, you can join Simplilearn's online agile training and improve your PMI-ACP exam preparation towards attaining a PMI-ACP agile certification. You can also attend agile certification classroom training courses for better insights. Know the upcoming dates of PMI-ACP agile certification training for your city.

As organisations have moved away from top-down decision making approaches as the education and skills of their employees have increased, the techniques used for project management have also evolved. Two major changes in organisational approaches to project management are generally grouped under the banners of "agile" and "lean" methods. An agile method relies upon incremental and iterative completion of goals with a self-managing team. It is often presented in opposition to a "waterfall" process that sequentially gathers requirements, completes a design, and then builds a final product.

Hirotaka Takeuchi and Ikujiro Nonaka proposed the core agile concept of iterative, continuous delivery in 1986. They are acknowledged as the inspiration for Scrum, a popular methodology for delivering IT projects today. Co-created by Ken Schwaber, Jeff Sutherland, John Scumniotales and Jeff McKenna, the term "Scrum" is often used interchangeably with "agile". Properly speaking, "Scrum" is a specific methodology whereas "agile" can be any technique that focuses on iterative delivery and empowerment. Agile primarily focuses on efficiently segmenting the business processing cycle of the problem-solving pattern into "chunks" that can be executed in parallel.

A lean method is one that aims to provide "perfect value to the customer through a perfect value creation process that has zero waste". Taiichi Ohno, developer of Kanban at Toyota, came up with the foundations of just-in-time planning and delivery mechanics such as Kanban that are at the core of lean.

In contrast to agile, which splits up work, lean focuses on continuous improvement of cross-functional teams and end-to-end process management. Lean places a far greater emphasis on the knowledge processing cycle of problem solving, continuously seeking new solutions to the status quo.

It is interesting to note the recent emergence of "Lean Six Sigma" advocates; this marks the reclamation of Lean philosophy by top-down management, instead of the more radical Toyota perspective of empowered, trusted, and self-managing teams6. But more importantly, following these rules places a relentless focus on discovering new problems, and in continuously building and enhancing the shared context required to effectively solve these problems.

The benefits of using problem solving methods to increase performance and adaptability have been repeatedly demonstrated. In fact, recent research suggests that the process of defining a problem as a series of small projects is often more important than picking a particular execution methodology

The effective problem-solving steps are the following:

- Identifying the issues: this entails being clear about what the problem is.
- Have an understanding of everyone's interest: this bring about the unity of purpose of all and sundry in the task of solving the problem. Allow everyone to make his or her contributions in the process of finding a solution to the problem.
- List the possible solutions: this requires time for brainstorming in order to creatively list the possible solution.
- Evaluation of the options: here, you will subject the options to a thorough evaluation.

- Selection of useful options.
- Documentation of the agreement: it entails writing down what is agreed upon by the stakeholders.
- Agree on contingencies, monitoring, and evaluation: this is done because there could be a change in condition. You should strive to make contingency agreements about foreseeable future circumstances, how you will monitor compliance and follow-through, and create opportunities for the evaluation of the agreements and their implementation.

Problem-Solving Strategies

A problem-solving strategy is a plan of action used in finding solutions to a problem. In solving a problem, different strategies have different action plans with them. The problem-solving strategies are:

Trial and error: this involves the use of different solutions until you are able to find a solution to the problem.

Algorithm: this is a problem-strategy that offers step-by-step instructions on how the problem can be solved to attain the desired outcome.

Heuristic: it is a general problem-solving framework. It includes the use of the "rule of thumb" which saves you time and energy when making a decision.

How to Solve Problems

In reality, the way you solve your personal or organizational problem will determine your success and happiness. The process of solving a problem includes defining the problem and breaking it into smaller pieces. Also, decide on how to approach the problem.

- You should endeavour to creatively approach the problem by working with other people in order to approach the problem from a different perspective. The following steps help you in solving a problem:
- Define the problem
- Make important decisions first: here, you will recognize the decisions that you need to make and how they will contribute to the task of solving the problem.
- Simplify the problem: this entails breaking the problem into smaller bits.
- Make an outline of what you know and what you don't know
- Be anticipatory about future outcomes: this assists you in coming up with different plans for different problems in the future.
- Allocate resources.

It is crucial for business organizations to have a problem-solving method and model in their workplace because having it enables them to have business problem-solving tools in their arsenal. Problem-solving method and models are used to address the challenges that arise in the workplace. It allows them to solve complex problems confronting their teams with a shared collaborative and systematic approach.

Also, the stakeholders that are involved in running an organization must have creative problem-solving skills at their disposal. This will go a long way in assisting them in confronting their business challenges.

A typical creative problem-solving example was used by XYZ Company, a vegetable oil company. The company was struggling to expand its business operations but the management members were able to creatively solve this problem by selling the shares of the company. With this, the company was able to raise funds required for its expansion.

Problem-solving and decision-making are vital skills for business and life. No doubt, problem-solving involves decision-making and decision-making is crucial for management and leadership. Hence, it is very paramount for business organizations and individuals to have good problem-solving and decision-making skills in order to fulfill their purpose. They also aid in gathering every fact thereby enhancing the understanding of the causes of the problems.

Work problem solving is the method of processing solutions to the work problem of a firm. It involves the collaborative efforts of the workers to solve problems and by doing so; the workers will contribute properly to the productivity of the organization.

A problem-solving flowchart is a diagram that gives a description of how an organization solves its problems. It gives a visualization of the process used in solving organizational problems. It can be used as a guide and reference for the future thereby saving the organizations a lot of time and energy.

Also, a problem-solving flowchart ensures teamwork and collaboration among the workers in confronting challenges they are faced with.

Problem management techniques enable an organization to be responsible for managing the lifecycle of all problems that could arise in an organization. It allows them to prevent problems and resulting incidents from happening, to estimate recurring incidents, and to minimize the impact of incidents that cannot be prevented.

Problem-solving in management is a key feature that any organization must have because it helps in identifying, analysing, and solving the problems that could hinder the growth of the business of the organization.

Also, it is crucial for a business firm to have in place problem-solving leadership that inspires and motivates the workforce to be creative in solving the problems that they will be faced with. Furthermore, problem-solving leadership creates an environment that allows the workers to opine their views on the problems of their organizations.

Agile Team Motivation

If a team want to achieve something they will find a way to make it happen. Conversely if the team don't want to achieve a goal, there is no way they will. Therefore, the single biggest factor which contributes to productivity is the motivation or morale of a team. Treating employees like volunteers means engaging with them and making sure they want to do what they have been asked to do. It would be better yet let them select what task they do Celebrate the success of every launch/release, even if only in a small way, but draw attention to the fact that the team were successful. Allow everyone to feel great about what they have achieved. Praise team members and offer some gold cards like compensatory off.

Traditional thinking is that you need to use a combination of rewards and punishment to motivate people who would otherwise be unmotivated.

Agile Failure Modes

Most of time team is struggling to get enough shared understanding to deliver a working tested increment of product every couple of weeks. With Scrum, we know we need clarity, accountability, and the ability to measure progress on frequent intervals, Once Scrum starts to go mainstream, all people remember the rules but they forgot the meaning behind the rules. Common failure areas are:

Culture doesn't support change - Try to keep cross-organizational uniformity and use PMO as enforcers

Lack of retrospectives Action items generally get lost, hence they should be tracked in system with due date assigned, and should be tracked in reports

Lack of collaboration in planning – Lack of communication and hence lack of collaboration

PMO or scrum master should force to sit in one common room during planning week, where they are forced to collaborate and communicate

Tsunami of technical debt - Stop and clear, one should not process till WIP or technical debt reach to level where it can be controlled

Chapter Eight

Conclusion

Software development methodologies have advanced since business requirements became more demanding. Agile methodologies came into existence after the need for a light way to do software development in order to accommodate changing requirements environment. Agile methodologies rules and practices require communication between the developer and the customer. Under pressure to stick and adhere to the Agile Methodologies principles and best practices, developers must be ready for any change at any time, while also having to maximize Stakeholders Investments. The main aim of agile methodologies is to deliver what is needed when it is needed and nothing more.

Agile Methodologies include a set of software development approaches. They have some variations, but still they share the same basic concepts. The main agile methodologies that are being used today are Extreme Programming (XP), Agile Modeling, and SCRUM. Extreme Programming (XP) is the coding of what the customer specifies, and the testing of that code. Agile Modeling defines a collection of values, principles, and practices which describe how to streamline modeling and documentation efforts. SCRUM is an Agile Methodology framework structured to support complex product development. Scrum consists of Scrum Teams and their associated roles, events, artifacts, and rules. Each component within the framework serves a specific purpose and is essential to Scrum's success and usage. Scrum is a simple low overhead process for managing and tracking software development. It attempts to control this 'chaordic' process using a project management framework that involves requirements gathering, design and programming.

Agile methodologies are not best suited for all projects. When communication between the developer and the customer is difficult, or when the development team does not have experienced developers, Agile Methodologies will not give the best results. These methodologies exhibit optimum results when there is a strong communication between the developer and the customer, and the development team compromises skilled team members. When there is a chance for misunderstanding the exact customer requirements, or when the deadlines and budgets are tight, then Agile methodologies are the optimum approach for a solution.

Agile is a way of thinking about how a software development can be managed. Regardless of the exact frameworks and techniques they use, 98% companies have realized success from Agile projects. Higher speed, flexibility, and productivity achieved through such approaches are the key drivers which motivate more and more organizations to switch to Agile. Software engineering, being an extremely fast-paced industry, calls for flexibility and responsiveness in every aspect of project development. Agile frameworks allow for delivering cutting-edge products and cultivating innovative experiences while keeping the product in sync with the market trends and user requirements.

However, there is always a place for diversity. Depending on your business requirements and goals, you might still benefit from using the Waterfall model or the combination of the two. Specific challenges with using an Agile Method can be offset by adding back some formality. Agile Methods offer software project managers an alternative developement and management methodology that provides good support for projects with ill defined or rapidly changing requiremnts. Even on project that are questionable for the application of the agie method, underlying agile principles may still be effective. Project Managers should consider its usage for such projects assuming that they have a team capable of using it and can implement the required processes.

www.ingramcontent.com/pod-product-compliance
Lightning Source LLC
Chambersburg PA
CBHW070657220526
45466CB00001B/475